Understanding the Elements of the Periodic Table™

RADON

Janey Levy

rosen publishing's
rosen central
New York

Published in 2009 by The Rosen Publishing Group, Inc.
29 East 21st Street, New York, NY 10010

First Edition

Library of Congress Cataloging-in-Publication Data

Levy, Janey.
Radon / Janey Levy.—1st ed.
p. cm.—(Understanding the elements of the periodic table)
Includes bibliographical references and index.
ISBN-13: 978-1-4358-5069-9 (library binding)
1. Radon. I. Title.
QD181.R6L48 2009
546'.756—dc22

2008014947

Manufactured in the United States of America

On the cover: Radon's square on the periodic table of elements. Inset: The atomic structure of radon.

Contents

Introduction

The gaseous element radon (Rn) gets attention largely because it's radioactive. You're probably familiar with the concept of radioactivity. News reports sometimes mention it. Movies have featured—and sometimes misrepresented—the consequences of exposure to radioactivity. But what do you really know about radioactivity?

Radioactivity is actually common in nature. All elements heavier than bismuth (Bi) are radioactive. They give off energy in the form of tiny bits of matter as the nuclei of their atoms decay, or break down. People have harnessed this energy for many purposes. Doctors use it to treat diseases such as cancer. Nuclear power plants use uranium (U) and plutonium (Pu) to produce electricity. However, the movies are at least partly right: radioactivity can cause great harm to living organisms. So, it must be used carefully. People who work with radioactive materials are checked regularly to make sure they're not receiving too much exposure.

At the Limerick nuclear power plant near Pottstown, Pennsylvania, radiation detectors at the plant entrance measured workers for radioactive contamination. Employees who had been exposed to high levels of radiation could be identified and decontaminated. This procedure helped protect the workers' health and helped prevent contamination of nearby residents and the environment. It was because of the detectors that Stanley Watras's contamination was discovered.

Radon that was given off by uranium in underground granite contaminated Stanley Watras's house, pictured here in Pottstown, Pennsylvania. Testing revealed that almost half the homes in the area had dangerously high levels of radon.

Watras was a construction engineer working on building the Limerick plant. In 1984, he kept setting off the alarms on the detectors. He set off the alarms every day for two weeks. Very worrying to Watras was the fact that the detectors showed he was highly contaminated throughout his entire body. Strangely, though, he set off the alarms when he was *going into* the plant. It turned out that Watras's home had a high concentration of radon, and that was what had contaminated Watras—not anything at the plant.

The radon that contaminated Watras is the heaviest and only radioactive member of the group of elements called the noble or inert gases. People have found a wide range of uses for radon since its discovery around

1900. Some scientists use it to study the water cycle and the movement of air masses. Other scientists are trying to find ways to use radon to predict earthquakes and volcano eruptions, which would save lives. Since its discovery, people have found numerous medical applications for radon. It's one of several radioactive elements that have been used to treat cancer. Perhaps more surprisingly, for years, many people believed that radon had health benefits and intentionally exposed themselves to small amounts of it. Some people continue the practice today, and there is some evidence to support it. However, radon is now best known as one of the principal causes of lung cancer.

Chapter One Radon: The Radioactive Noble Gas

Radon's chemical symbol is Rn. Although it was discovered around 1900, it didn't receive the name "radon" until 1923. It was originally known by several different names. Scientists discovered that radium (Ra), thorium (Th), and actinium (Ac) each gave off a radioactive gas as it decayed and gave each gas its own name. Eventually, scientists recognized that the various gases were in fact the same gas. In 1923, the International Committee on Chemical Elements announced that the gas would officially be known as radon.

The "rad-" part of the name was taken from radium. There were several reasons the name honored radium, rather than thorium or actinium. The radioactive gas was first observed in the study of radium, radium gives off more of it than thorium or actinium, and the radon given off by radium lasts longer because it decays more slowly than that given off by thorium and actinium. The "-on" part of the name was chosen because it was the ending used for the other noble gas names (except for helium [He]).

The Discovery of Radon

The story of radon's discovery begins with the discovery of radioactivity. In 1895, German physicist Wilhelm Conrad Röntgen (1845–1923) discovered that he could create a picture on a photographic plate without exposing the

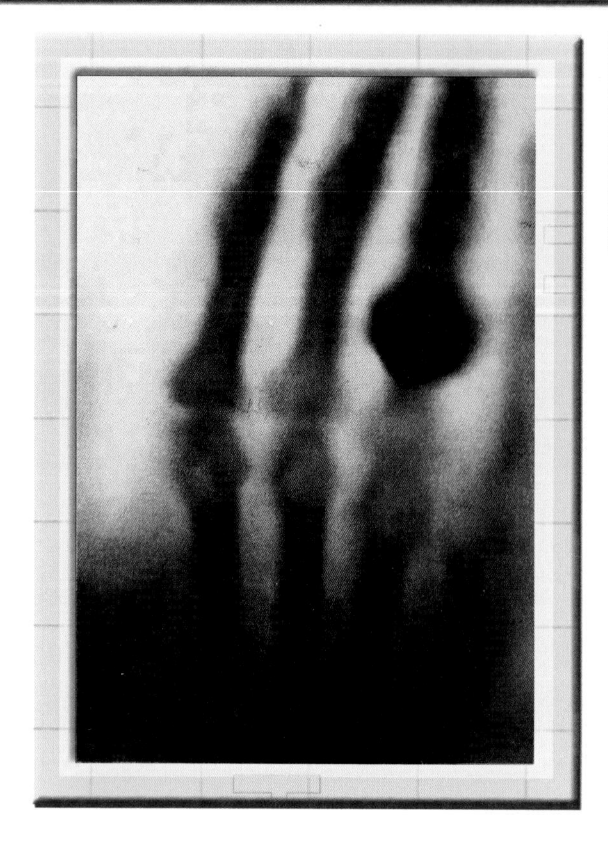

Wilhelm Röntgen's X-ray image of his wife's hand was hailed as one of humanity's greatest technological accomplishments. People predicted X-rays would revolutionize human life.

plate to light. He sealed a glass tube containing gas at a low pressure inside a black carton and sent an electric current through the tube. It produced invisible rays that made a picture of the bones in his wife's hand on the plate. Röntgen had discovered X-rays. This discovery earned him the first Nobel Prize in Physics in 1901.

French physicist Antoine Henri Becquerel (1852–1908) learned of Röntgen's experiments and began his own investigation in 1896. Becquerel discovered that uranium ore gave off invisible rays that created an image on a photographic plate. No electricity was used to produce these rays. The ore gave them off spontaneously. Becquerel had discovered radioactivity.

Becquerel suggested to young physicists Marie (1867–1934) and Pierre Curie (1859–1906) that they conduct experiments on radioactivity. They discovered that uranium ore generated more radiation than did pure uranium. That led to the discovery in 1898 of two new radioactive elements in the ore, which they named radium and polonium (Po). Together, Becquerel and the Curies received the 1903 Nobel Prize in Physics for their discoveries.

During their experiments, the Curies noticed the air surrounding radium became radioactive. They believed an immaterial radioactive energy given off by radium caused this effect. British scientists Ernest Rutherford

Ernest Rutherford's work brought him great recognition. He was knighted in 1914 and became president of the Royal Society of London in 1925, the year this photograph was taken.

(1871–1937) and Frederick Soddy (1877–1956) began a thorough investigation of the phenomenon in 1900. Their experiments showed that the "immaterial radioactive energy," or emanation, was actually a radioactive gas. They believed it was a new element in the recently discovered family of noble gases.

Other scientists were also investigating radioactivity. French chemist André-Louis Debierne (1874–1949) discovered actinium in 1899 and found that it had an emanation. Inspired by Rutherford and Soddy's work, German chemist Friedrich Ernst Dorn (1848–1916) conducted his own experiments in 1900. He found that both thorium and radium produced an emanation but did not study it any further.

In 1910, Scottish chemist Sir William Ramsay (1852–1916) and British chemist Robert Whytlaw-Gray (1877–1958) took the final steps in identifying radon. They proved it was a noble gas and established its place in the periodic table of elements.

As this story shows, the discovery of radon wasn't a single event. It was a series of events involving many scientists. Yet, we still usually give one scientist credit for the "big breakthrough." For years, Dorn was credited with radon's discovery. In 2003, however, researchers James L. and Virginia R. Marshall showed that the credit really belonged to Rutherford.

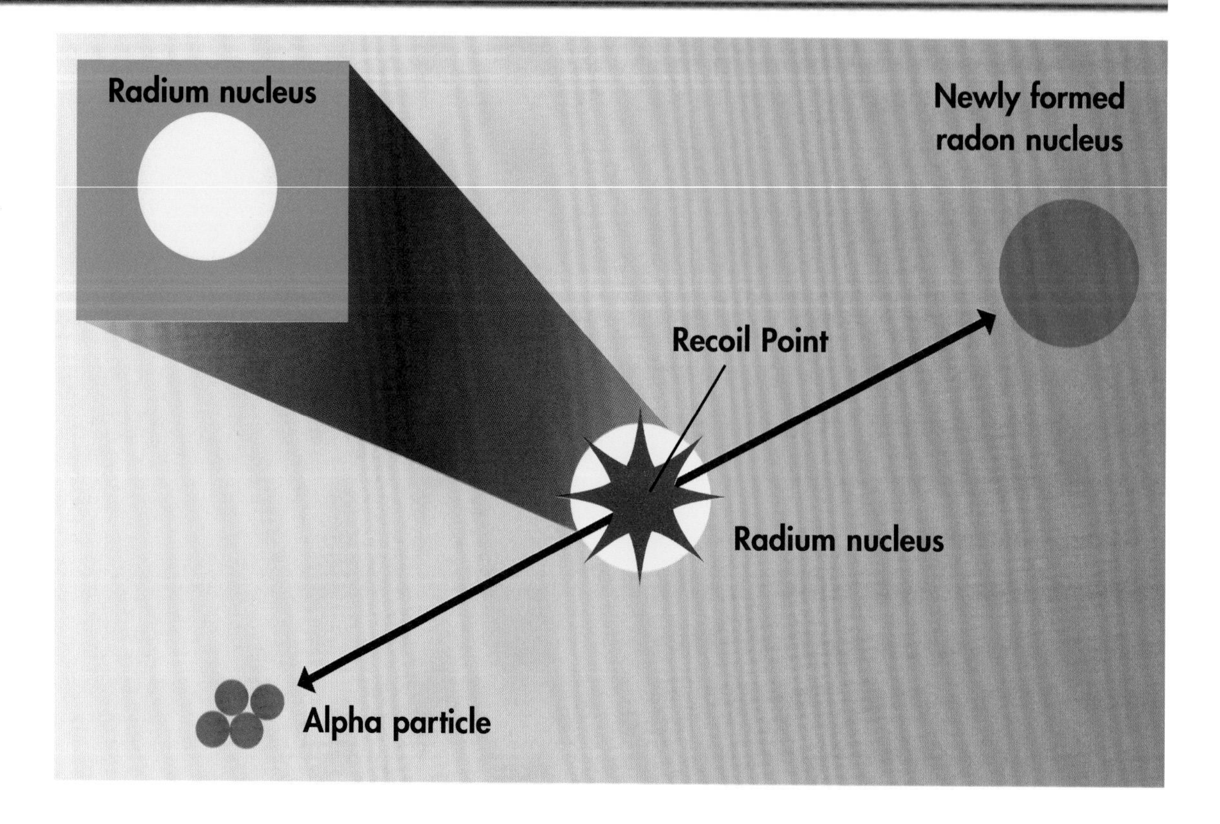

When a radium nucleus decays, it sends out an alpha particle in one direction. The newly formed radon nucleus recoils, or springs back, in the opposite direction.

Organizing the Elements

By the time radon was discovered, scientists had identified more than eighty elements. How did they keep all the information straight? How did they make sense of the relationships between elements? The need for some system of organization had become clear by the early 1800s. In 1817, German chemist Johann Dobereiner (1780–1849) made one of the earliest attempts to organize the elements. Others followed. By the 1860s, several scientists were trying to create organization systems for the elements.

French geologist A. E. Beguyer de Chancourtois (1820–1886) created a periodic table in 1862 that arranged the known elements in order of increasing atomic weight (expressed as atomic mass units, or amu).

This handwritten paper, which is dated February 17, 1869, at the lower left, shows Dmitry Mendeleyev's first version of his periodic table. The many changes indicate that Mendeleyev was still working out his ideas.

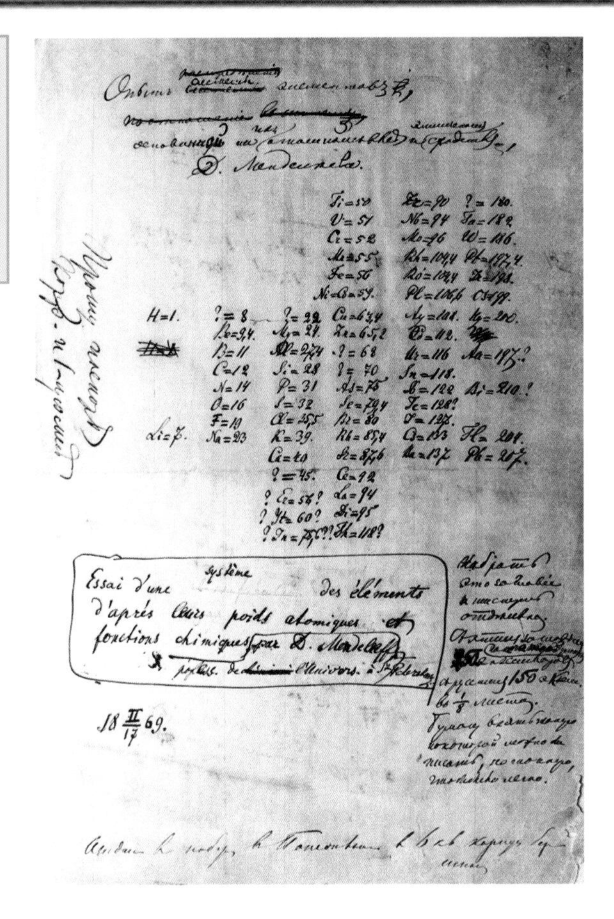
Essai d'une système des éléments d'après leurs poids atomiques et fonctions chimiques par D. Mendeleeff

18 II/17 69.

However, it looked quite different from the modern table. Around 1865, British chemist John Newlands (1837–1898) discovered that when the elements were arranged in order of increasing atomic weight, a pattern emerged in which their properties repeated periodically, or at regular intervals. Every eighth element in Newlands's table had similar chemical properties. This interval reminded him of a musical scale, so he called the pattern the law of octaves. As new elements were discovered, it quickly became clear that this precise interval didn't always work. But Newlands had recognized the basic feature of the periodic law.

The periodic law states that if elements are arranged in order of increasing atomic weight, elements with similar properties appear at regular intervals. Russian chemist Dmitry Mendeleyev (1834–1907) and German chemist Julius Lothar Meyer (1830–1895) independently discovered the law in the late 1860s. However, Mendeleyev announced his discovery first, so he gets the credit.

In 1869, Mendeleyev arranged the known elements in a table based on the periodic law. He arranged elements in each row, or period, from left to right in order of increasing atomic weight. When he came to an

What's in a Name?

Radon went by many names before scientists settled on "radon." Because it emanates or flows from another element, it was originally called emanation. This name comes from the Latin word *emanare*, which means "to flow out." Radon was also known as niton, from the Latin *nitens*, which means "shining." The gas released by radium was sometimes called radeon. Thorium's gas was called thoron or thoreon. Actinium's gas was called akton, actineon, or actinon.

element that had properties similar to the element at the left end of the row, he started a new row. The result was that the elements in each column, or group, had similar properties. The known elements of Mendeleyev's time did not completely fill the table. Mendeleyev correctly believed elements would be discovered to fill the gaps and was even accurately able to predict the properties of many still-undiscovered elements.

Mendeleyev's table formed the basis of the periodic table that people use today. However, it had far fewer elements than the modern table. Among the missing elements were the noble gases because none had yet been discovered. The first was discovered in 1894. Working independently, Sir William Ramsay and British physicist Lord Rayleigh (John William Strutt) (1842–1919) discovered argon (Ar) that year. Ramsay went on to discover helium in 1895, and krypton (Kr), neon (Ne), and xenon (Xe) in 1898. He received the 1904 Nobel Prize in Chemistry for his work. Given Ramsay's role in the discovery of the noble gases, it's not surprising he was the one who provided official proof that radon was a noble gas and put it in its proper place in the periodic table.

Chapter Two
Radon, Atomic Structure, and the Periodic Table

You can't see them, but atoms are everywhere. They make up everything. Scientists have discovered more than one hundred kinds. Each element is composed of one kind of atom. So, what exactly are atoms? Atoms are extremely tiny bits of matter. They're so tiny that you would have to place about 300,000 radon atoms side by side to equal the width of a single human hair! Yet, atoms aren't the smallest bits of matter. They're made up of tinier bits called subatomic particles.

Atomic Structure

Atoms are sometimes said to resemble tiny solar systems. A heavy center object called the nucleus provides almost all the atom's mass. Neutrons, which have no electrical charge, and positively charged protons make up the nucleus. These particles give the nucleus an overall positive charge.

Negatively charged, lightweight particles called electrons orbit the nucleus. The number of electrons usually equals the number of protons, which means the atom as a whole has no electrical charge.

The electrons occupy overlapping shells, or energy levels, around the nucleus. Only a certain number of electrons can be in each shell. Two can occupy the first shell, eight can occupy the second, eighteen can occupy the third, and thirty-two can occupy the fourth. In theory, many more

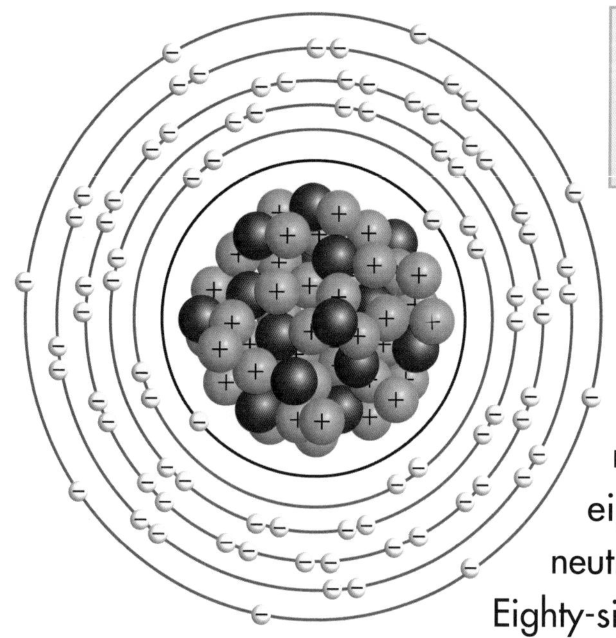

Radon has two electrons in its first shell, eight in the second shell, eighteen in the third, thirty-two in the fourth, eighteen in the fifth, and eight in the sixth.

electrons can occupy each of the three remaining shells. In the real world, however, these shells are never full. All radon atoms contain eighty-six protons. Most have 136 neutrons, but some have 134 or 133. Eighty-six electrons, arranged in six shells, orbit the nucleus.

How Altering Atomic Structure Changes Elements

What happens if you fiddle with an atom's structure? Different changes have different effects. Gaining or losing an electron produces an ion. It doesn't change what element an atom is—the number of protons determines that. It simply gives the atom an electrical charge. Gaining an electron creates a negatively charged ion. Losing one creates a positively charged ion.

Because the number of protons governs what element an atom is, changing that number transforms the atom. If you were to add a proton to the nucleus of a radon atom, you would have francium (Fr). Francium—like radon—is produced by the decay of radioactive elements and is itself radioactive. Unlike radon, it's a metal, not a gas. If a proton were subtracted from the nucleus of a radon atom, you would have astatine (At). Astatine is a solid, not a gas, at room temperature. It has some properties of metals, although it isn't one. Like both radon and francium, it's radioactive. The number of protons in an atom is known as the atomic number. Thus,

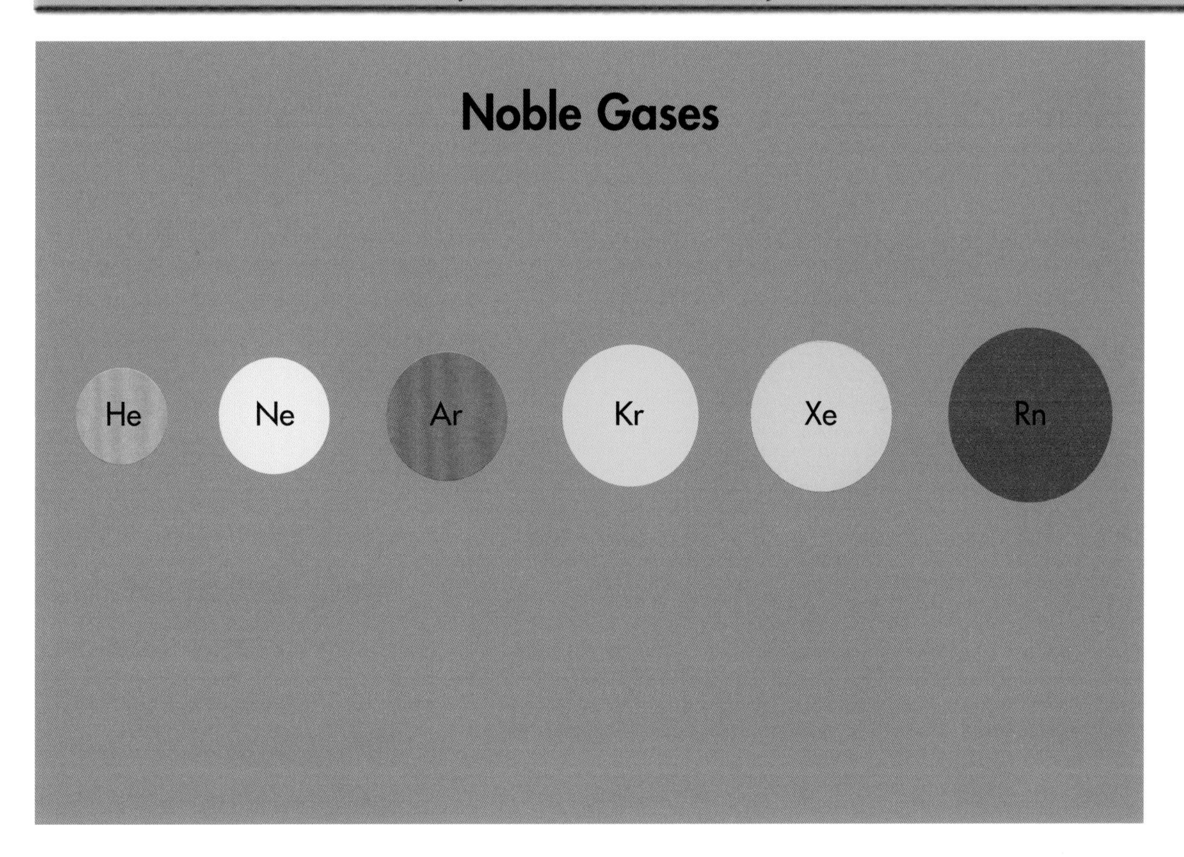

Of the noble gases, only radon is radioactive. That is, only radon spontaneously changes its atomic structure. Ernest Rutherford and Frederick Soddy applied the term "transmutation" to this property of radioactive elements.

radon, which has eighty-six protons, has an atomic number of 86. In the periodic table reproduced in this book (on pages 40–41), each element's atomic number appears in the upper left corner of the element's square.

Changing the number of neutrons changes the element's isotope. Most elements have several isotopes. Scientists usually refer to a particular isotope by the element's name plus the isotope's mass number, which is the number of protons plus the number of neutrons. Radon has three naturally occurring isotopes. The one that lasts the longest and occurs in the greatest quantities is radon-222, which has 86 protons and 136 neutrons (86 + 136 = 222). More than thirty isotopes of radon have been created in laboratories.

An element's square on the periodic table provides not only its chemical symbol, but also its atomic number and atomic weight. Using the two numbers, what can you determine about the atomic structure of the element?

Because the number of neutrons varies among isotopes, each isotope has a different atomic weight. Scientists determine an element's official atomic weight by averaging the atomic weights of its naturally occurring isotopes, taking into account the proportions in which they occur. Each element's atomic weight appears to the upper right of the element's symbol on the periodic table in this book. Radon's atomic weight is 222.0176, which has been rounded to 222 in our periodic table.

It Glows!

Radon is invisible at room temperature—a colorless, odorless, tasteless gas. It remains colorless when it becomes liquid at –79° Fahrenheit (–62° Celsius). However, at –96°F (–71°C), radon is transformed into a phosphorescent, or glowing, colorful solid. It gives off a soft yellow glow that deepens to orange-red as the radon is cooled to –319°F (–195°C).

Radon Snapshot

86 222
Rn

Chemical Symbol:	Rn
Classification:	Nonmetal; noble gas
Properties:	Nonreactive, radioactive, colorless, odorless, tasteless
Discovered By:	Ernest Rutherford in 1900
Atomic Number:	86
Atomic Weight:	222.0176 atomic mass units (amu), sometimes rounded to 222
Protons:	86
Electrons:	86
Neutrons:	136, 134, or 133 (in order of decreasing abundance)
State of Matter at 68°F (20°C):	Gas
Melting Point:	–96°F (–71°C)
Boiling Point:	–79°F (–62°C)
Commonly Found:	Not abundant in nature, but found in Earth's crust, bodies of water, and atmosphere

What the Periodic Table Tells You About Radon

An element's location on the periodic table tells you information about it. The modern periodic table arranges elements in order of increasing atomic number, making the patterns of the periodic law even clearer than they were in Mendeleyev's table. That means an element's place on the table tells you much about its properties even if you don't know anything about it.

An element's relation to the staircase line reveals whether it's a metal, nonmetal, or metalloid (an element that possesses some properties of both metals and nonmetals). The staircase line is on the right side of the table, where the colored blocks create a series of steps or stairs. Elements to the left of the staircase line are metals. Those to the right are nonmetals. Elements touching the line are metalloids. Radon is to the right of the line, so you know it's a nonmetal.

Nonmetals lack the distinctive properties of metals. Metals are usually solids that can be polished to a shine. They conduct electricity. They're also malleable and ductile. Malleable substances can be hammered into shapes without breaking. Ductile substances can be stretched into wires.

You'll also notice that the periodic table is organized into rows and columns. The rows, or periods, are numbered 1 through 7. Elements in each period have the same number of electron shells, and the period's number tells how many electron shells its elements have. Radon is located in period 6, and it has six electron shells.

The columns, or groups, are also numbered. There are two systems for numbering the eighteen groups. In the traditional system, Roman numerals are combined with letters of the alphabet. In the newer system, Arabic numerals are used. Generally, the elements in each group have the same number of electrons in their outer shell. These electrons, called valence electrons, have a lot to do with an element's chemical behavior. Each

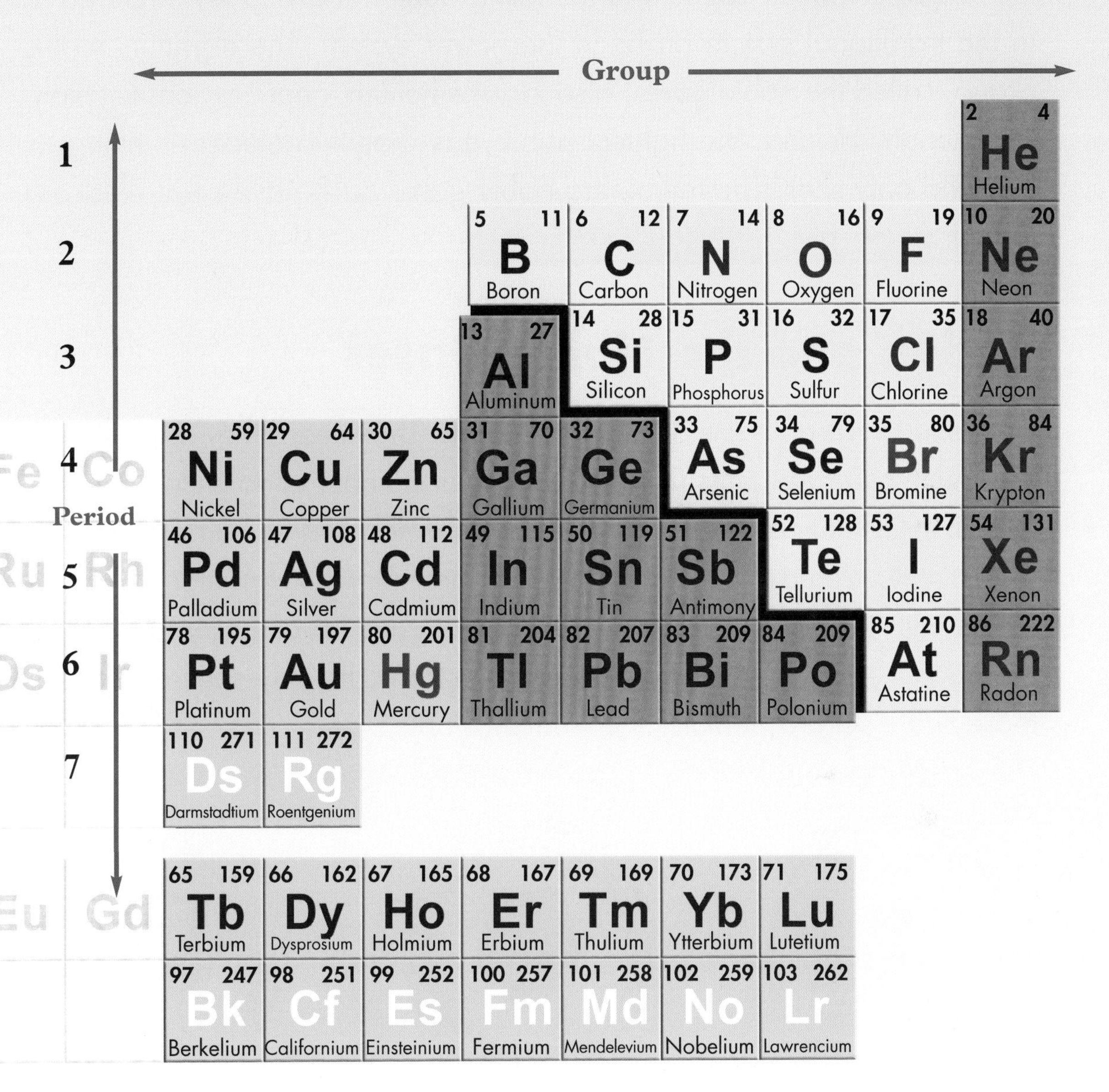

The periodic table sorts elements into groups and periods. The right side of the periodic table is pictured here. Radon's group and period numbers help you locate it on the table. Find its group number across the top. Then find its period number on the left side.

group's elements behave similarly because they (usually) have the same number of valence electrons.

Radon is the last element in the last group. The group is numbered O in the traditional system or 18 in the newer system. The elements in this group, called the noble gases, also include helium, neon, argon, krypton, and xenon. Helium, the lightest noble gas, has two electrons filling its one and only shell. The remaining noble gases each have eight electrons in their outer shell.

Radon and the Noble Gases

The noble gases are the most stable elements. That means they're unlikely to react with other elements. This is due to their number of valence electrons. Except for helium, all noble gases have eight electrons in their outer shells. This is considered the most stable arrangement for an atom because atoms with this arrangement are unlikely to give up or accept electrons. Although helium has just two valence electrons, its only shell is full with those two electrons, making the element stable.

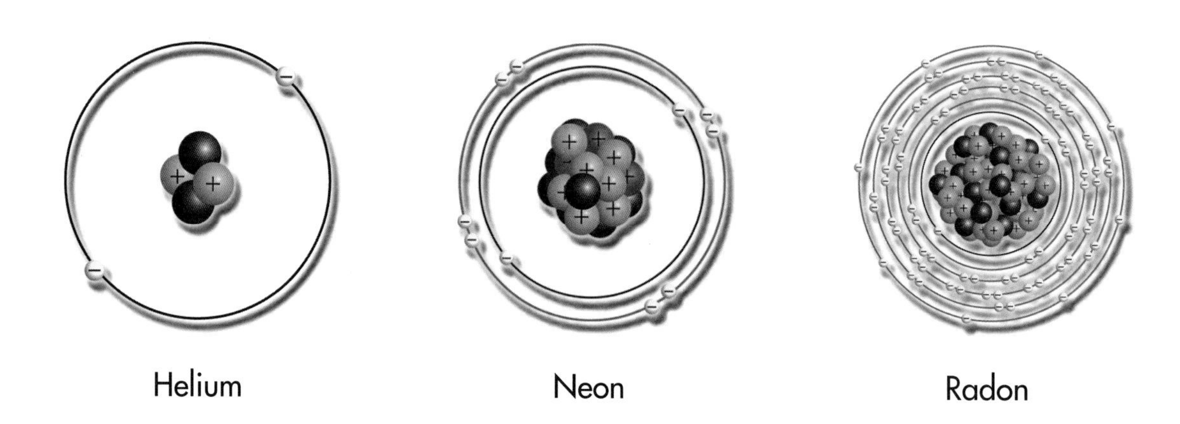

Although the noble gases all behave similarly, there are important differences among them. This illustration compares the atoms of the two smallest and lightest noble gases with radon's, the largest and heaviest in the group.

The noble gases have been known by various names. They were believed to be rare when they were discovered, so they were called the rare gases. However, scientists learned that some are not rare. Scientists have also called them inert gases because they are generally nonreactive. Yet, there are problems with this name, too. The term "inert" has other meanings in chemistry. Gases outside this group—such as nitrogen (N) and carbon dioxide (CO_2)—are called inert to indicate that they're not flammable. In addition, compounds of most noble gases have been made. An element that can form compounds is not totally nonreactive. Finally, the term "noble" was applied to the group. For centuries, scientists have used the term to describe substances that don't react with oxygen. These gases don't react with oxygen, so "noble" is a good name for them.

Chapter Three
Radon: A Radioactive Product of Radioactive Decay

Radon is found almost everywhere on Earth in trace amounts. It's in the planet's crust, in bodies of water, and in the atmosphere. Because it's a radioactive element produced by the decay of other radioactive elements, it's constantly being created and destroyed. That's because the process of radioactive decay transforms one element into another. Radioactive decay often changes the number of protons in an atom's nucleus, and, as noted earlier, changing the number of protons changes what element an atom is. So, the decay of heavier radioactive elements creates radon, and radon's own decay destroys the element and creates lighter elements.

Radon-222 is the most common isotope of radon. It's also the most stable, as it has the longest half-life. The term "half-life" refers to the rate at which radioactive elements decay. The half-life is the time it takes for half the atoms in a sample of an element to decay. The half-life of radon-222 is 3.82 days, which isn't long compared to other radioactive elements found in nature. The half-life of radium-226, for example, is 1,630 years. The half-life of uranium-238 is 4.6 billion years!

Radon-222 is the product of the radioactive decay of radium-226, which is itself the product of the radioactive decay of uranium-238. Thus radon-222 belongs to the radioactive decay series known as the uranium series.

This diagram of the radium-226 decay chain shows whether the changes involve alpha or beta decay. The word "gamma" indicates changes that also involve giving off high-energy gamma rays.

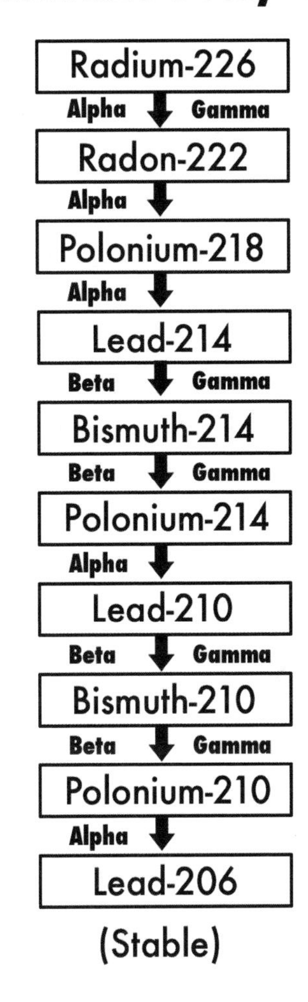

How Radioactive Decay Works

Elements that are naturally radioactive decay through particle radiation. That is, they decay by giving off electrically charged particles called alpha particles or beta particles. An alpha particle consists of two protons and two neutrons, and has a positive charge. A beta particle consists of an electron that is released when a neutron changes into a proton. Because beta particles are electrons, they are negatively charged. Alpha decay reduces the number of protons in the nucleus. Beta decay increases the number of protons.

During an alpha or beta decay process, the original element, or parent, is transformed into a different element, called a daughter element, or progeny. Ernest Rutherford and his assistant, Frederick Soddy, uncovered this phenomenon, which they called transmutation. They published their discovery in 1902. The work earned Rutherford the 1908 Nobel Prize in Chemistry.

The progeny produced by nuclear decay can also be radioactive. When it is, it produces its own progeny. This decay process is repeated

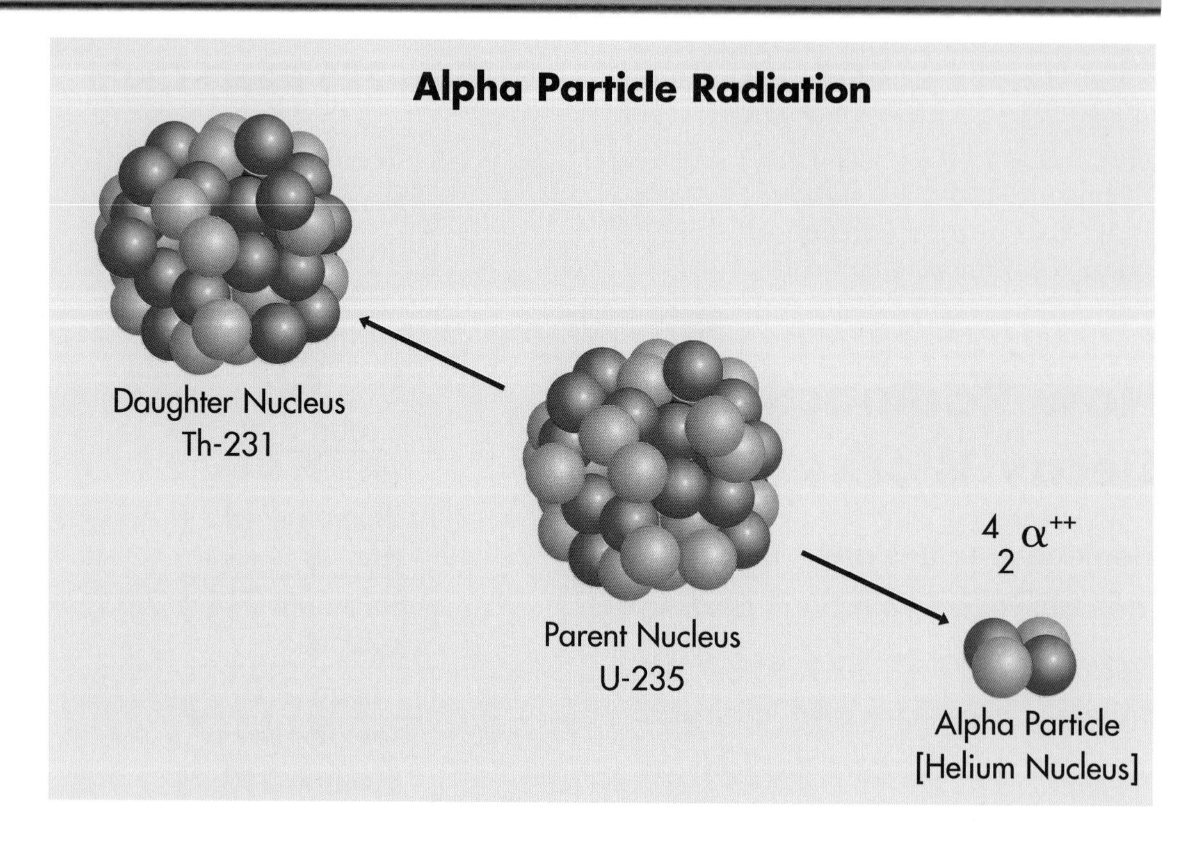

This diagram illustrates the alpha decay process of a uranium-235 nucleus. Notice it says "helium nucleus" below the alpha particle. Compare this figure to the helium atom that is pictured on page 20.

until a stable isotope—one that isn't radioactive—is formed. The sequence of isotopes formed by successive decays is called a radioactive decay series. Uranium-238 is the parent of the uranium series. Uranium-238 occurs in most rocks and soils, which means its progeny also do. Radon-222 is one of the progeny in this series.

The other two naturally occurring radon isotopes are produced by the two other decay series found in nature—the thorium series and the actinium series. These radon isotopes are difficult to study because of their short half-lives. The thorium series produces radon-220, which has a half-life of 54.5 seconds. The actinium series produces radon-219, which has a very short half-life of only 3.92 seconds!

Marie Curie

Marie Curie, born in 1867, was one of the leading scientists of her time. Although Antoine Henri Becquerel discovered radioactivity, Marie Curie was the one who named it. She was also the first to recognize it was a property of individual atoms rather than molecules. She was the first woman to receive a Nobel Prize and the first person to receive two (1903 and 1911). Curie died in 1934, the victim of long years of exposure to radiation.

Marie Curie is shown here at work in her laboratory in 1910, seven years after she received her first Nobel Prize in Physics.

Radon's Progeny

Because radon is radioactive, it doesn't mark the end of any of the decay series. Radon produces its own progeny, and the decay continues until it produces an element that is stable.

The same elements form radon's progeny in all three decay series, although each series has its own isotopes of the elements. Those elements are polonium, lead (Pb), astatine, bismuth, and thallium (Tl). In all three decay series, radon's first daughter is a polonium isotope, and the final

Ernest Rutherford

Ernest Rutherford, born in 1871 in New Zealand, is considered the father of nuclear science. In addition to discovering transmutation, he invented the names "alpha particle" and "beta particle" for the particles given off during radioactive decay. He also created the nuclear model of the atom. Around 1919, he became the first person to break up the nucleus of an atom. He did his research in radioactivity in both England and Canada. In recognition of his achievements, Rutherford was made a British baron in 1931. He died in 1937.

element in the series is a stable isotope of lead. The uranium series ends with lead-206, the actinium series with lead-207, and the thorium series with lead-208.

Although radon is a gas, all its progeny are solid elements. As you will discover in the next chapter, this characteristic makes radon especially dangerous.

Chapter Four
Uses and Dangers of Radon

Radon's radioactivity is the sort of thing that people often call a double-edged sword—it's both good and bad. The radioactivity makes radon very useful for certain purposes. However, being around radon can put people's health at risk. In fact, radon is believed to be one of the leading causes of lung cancer. So, any decision to use radon for any purpose must be carefully weighed with the possible benefits against the serious risks.

Scientific and Industrial Uses of Radon

Scientists are exploring ways to use radon to learn more about Earth. Radon's widespread presence in Earth's crust makes it convenient for such scientific studies. Its radioactivity makes it easy to detect. Its gaseous state and its behavior—the way it dissolves easily in water and the way it moves through soil and air—make it a promising tool for studying several natural phenomena.

Some scientists are measuring radon in streams to help better understand aspects of the water cycle. Radon levels may be high in groundwater but not usually in streams because radon in a stream quickly enters the atmosphere. Scientists can identify places where groundwater enters a stream by locating places where the stream has high radon concentrations.

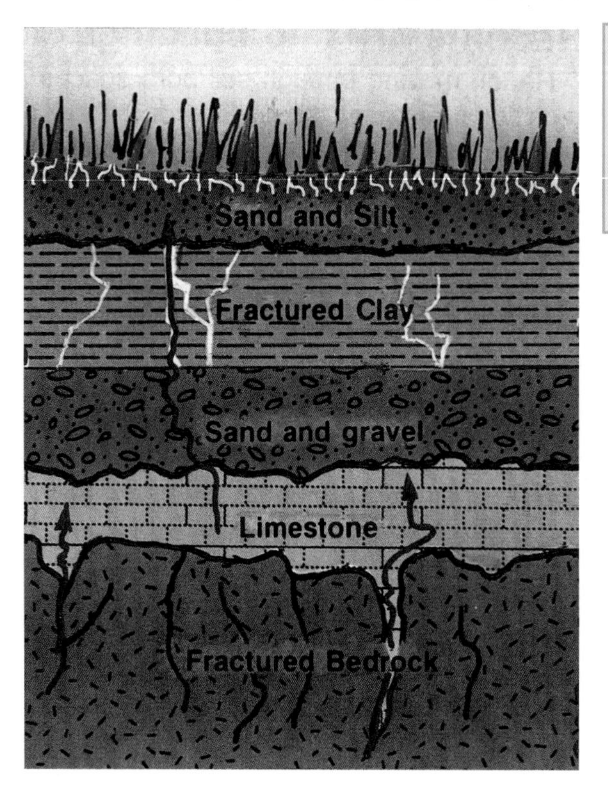

Gaseous radon moves upward easily through soil, sand, and gravel. It moves slowly through rock, but if there are any cracks or fractures radon will move upward more quickly.

Some scientists use radon to track the movement of air masses. They can do that because the amount of radon entering the atmosphere varies from place to place according to the type of soil and how much uranium is in the ground.

Other scientists hope to use radon to predict earthquakes and volcano eruptions. One method takes advantage of the fact that radon dissolves easily in water. Scientists have found that elevated radon levels often occur in wells and groundwater before an earthquake. However, scientists have also found that elevated levels may occur during or after an earthquake rather than before. Another method involves measuring radon levels in soil and in air close to the ground. As a gas, radon moves upward through soil and into the air. Its movement through soil is influenced by factors such as the type of soil, the amount of moisture in the soil, and the number of cracks and breaks in underground rocks. That means the faults in Earth's crust—places where many volcanoes are located and where earthquakes often happen—affect concentrations of radon in soil and air above them. Significant variations in radon concentrations may occur along faults around the time an earthquake happens or a volcano erupts. Scientists hope to learn enough about these variations to use them to predict earthquakes and volcano eruptions.

Large surface cracks may appear in the ground after an earthquake, such as this one in California in 1989. Such cracks may increase radon concentrations in the air.

Radon has some industrial uses as well. Because it's part of the uranium decay series, its presence in soil or in air near the ground is an indication that uranium is present. So, industries that explore for uranium can find it by detecting radon. In a similar way, oil companies can locate petroleum by measuring for radon. That's because radon dissolves well in oil-like substances.

Radon is also sometimes used to check for flaws in products. For example, radon can be added to gas or liquid flowing through a tube. If radiation is detected outside the tube, it indicates there's a leak. Radon's radioactivity can reveal flaws in metal castings, too. The flaws are visible in images that the radioactivity creates on photographic plates.

Medical Uses of Radon

The health dangers of radioactivity weren't recognized until many years after its discovery. This was partly due to the fact that the illnesses caused by radioactivity take years to develop. Because the dangers were not apparent, many people, including doctors, were excited by the discovery of radioactivity. They believed radioactivity might have health benefits. During the first half of the twentieth century, radon was used to treat a variety of health conditions.

Mineral springs, or spas, that are rich in radon enjoyed great popularity in the early twentieth century. Bathing in such springs, breathing the surrounding air with its high levels of radon, and even drinking the water were fashionable health treatments. People in the United States, Europe, and Japan believed these treatments could cure health problems such as arthritis, asthma, allergies, diabetes, and high blood pressure.

Belief in radon's health benefits was so strong that people wanted to take advantage of the gas's supposed positive effects, even if they couldn't visit a spa. So, radon was added to water, and the solution was then sold as a health tonic. The drink was widely favored in the 1920s and 1930s.

A father and his son enjoy bathing at Misasa, an extremely popular Japanese radon spa. Every year, tens of thousands of people visit radon spas and mines because of claims of supposed health cures.

Radon "seeds" were another popular health treatment. The "seeds" were small sealed gold or glass tubes filled with radon. As early as 1914, they were used to treat cancer. Doctors inserted the seeds directly into the cancer tumor in order to kill the cancer cells. Between 1930 and 1950, radon seeds were also used to treat acne and other skin conditions.

Today, radon is not widely used in medicine. Less expensive and safer isotopes of other elements are commonly used to treat cancer. Spas and mines with high levels of radon in the air have retained greater popularity. People still visit them seeking relief from numerous health problems. Japanese and Russian researchers have been studying the effects of spas and mines, and they claim to have found evidence that they do actually

Radon Compounds

Compounds are formed when two or more elements bond together. Radon and other noble gases don't form compounds in nature. However, compounds can be created in a laboratory, and scientists have made compounds of radon and the highly reactive element fluorine (F). No uses have been found for these compounds, and they haven't been studied much. Radon's radioactivity makes them dangerous. Also, radon's short half-life means they don't last long.

have some benefit. Most scientists, however, remain unconvinced. Moreover, many scientists point out that any benefits are likely to be outweighed by the demonstrated health dangers of radon exposure.

Dangers of Radon

The very radioactivity that makes radon useful also makes it dangerous. Only cigarette smoking causes more lung cancer than exposure to radon. The National Academy of Sciences (NAS) estimates that 15,000 to 22,000 people in the United States die every year from lung cancer caused by radon. You might be surprised to learn it's not the radon itself that presents the greatest danger. It's radon's radioactive progeny.

When you breathe in air containing radon, you breathe in radon. Yet it doesn't stay in your lungs for long. It goes out when you breathe out. However, as mentioned earlier, all radon's progeny are solid elements. They attach to dust and other particles in the air. They enter your lungs when you breathe in and can get stuck there, where they continue to give off radiation. The radiation kills cells or causes them to grow abnormally.

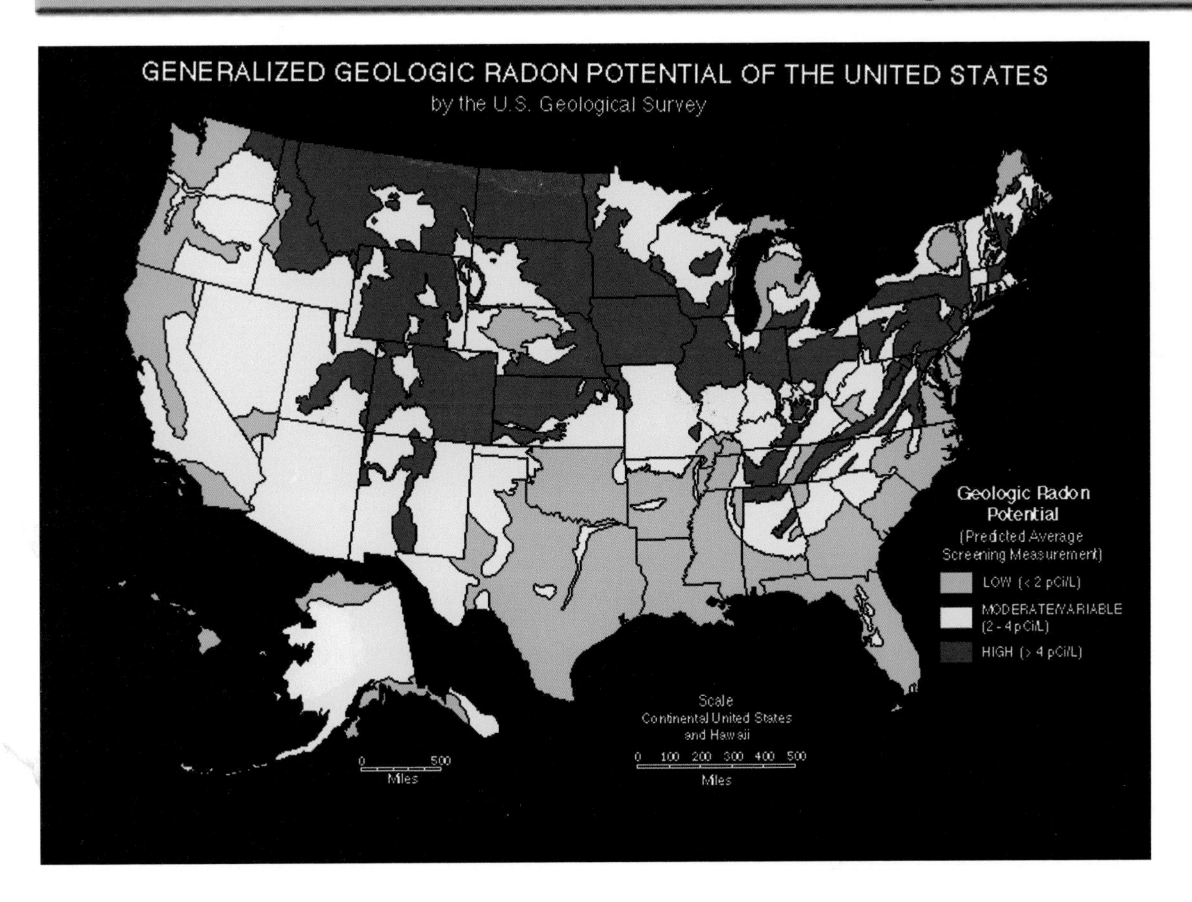

This U.S. Geological Survey map shows the level of risk from radon across the United States. Based on this map, is the risk high, low, or moderate where you live?

This can lead not only to lung cancer but also to emphysema and other lung diseases.

Because radon dissolves easily in water, it may also be in your drinking water. That can cause stomach and other forms of cancer. However, the number of deaths from these radon cancers is quite small. The NAS estimates only about eighteen people in the United States die each year from these forms of radon-caused cancer.

Chapter Five
Radon and You

By now, you've learned there's probably radon in the air where you live because it's found almost everywhere. Radon is constantly entering the air from the soil and groundwater. Is this a big problem? Should you be afraid to breathe when you're outside? The answer is no. Radon in the outside air isn't a serious danger. Once radon enters the atmosphere, it quickly diffuses through the air. As a result, concentrations are too low to pose a great risk. However, radon can be a problem in places you might not expect it, places where you feel safe. High levels can occur inside homes, schools, and other buildings. Fortunately, there are ways to fix the problem.

Stanley Watras's story alerted many people to radon's dangers. Two years later, this New Jersey family purchased a kit to reduce their home's radon levels.

Radioactive Cigarettes

You already know that cigarette smoke contains harmful tar and carbon monoxide (CO). But did you know that cigarettes are also radioactive? Radon progeny cling to the sticky leaves of tobacco plants. If you smoke, you breathe in the progeny. As you've learned in this book, the progeny stick in your lungs, where their radiation can cause lung cancer and emphysema. Remember this fact the next time you think about smoking—or breathe someone else's smoke.

Radon in Buildings

Your reaction to learning that radon can be a problem in homes may be to think that you should keep the windows closed to keep radon out. However, that will actually make the problem worse. Open windows aren't the main way radon enters buildings, and poor ventilation keeps the radon inside and makes the indoor concentrations higher. An unintended result of building structures that are better insulated and more energy efficient is that radon can get trapped inside them and build up to dangerous levels.

So, how does radon get inside buildings? It enters from the soil and rocks under the building. It's especially a problem in buildings with basements, but it can be a problem in other buildings as well. It gets in through cracks in basements, foundations, floors, and walls; construction joints; and gaps around pipes that carry water or gas into and out of the building. Radon can also get in through the water supply, especially if the water comes from a well. When water is agitated, as it is when you shower or wash clothes, it readily releases radon into the air. Occasionally, radon even enters a structure from the building materials used in it. Radon

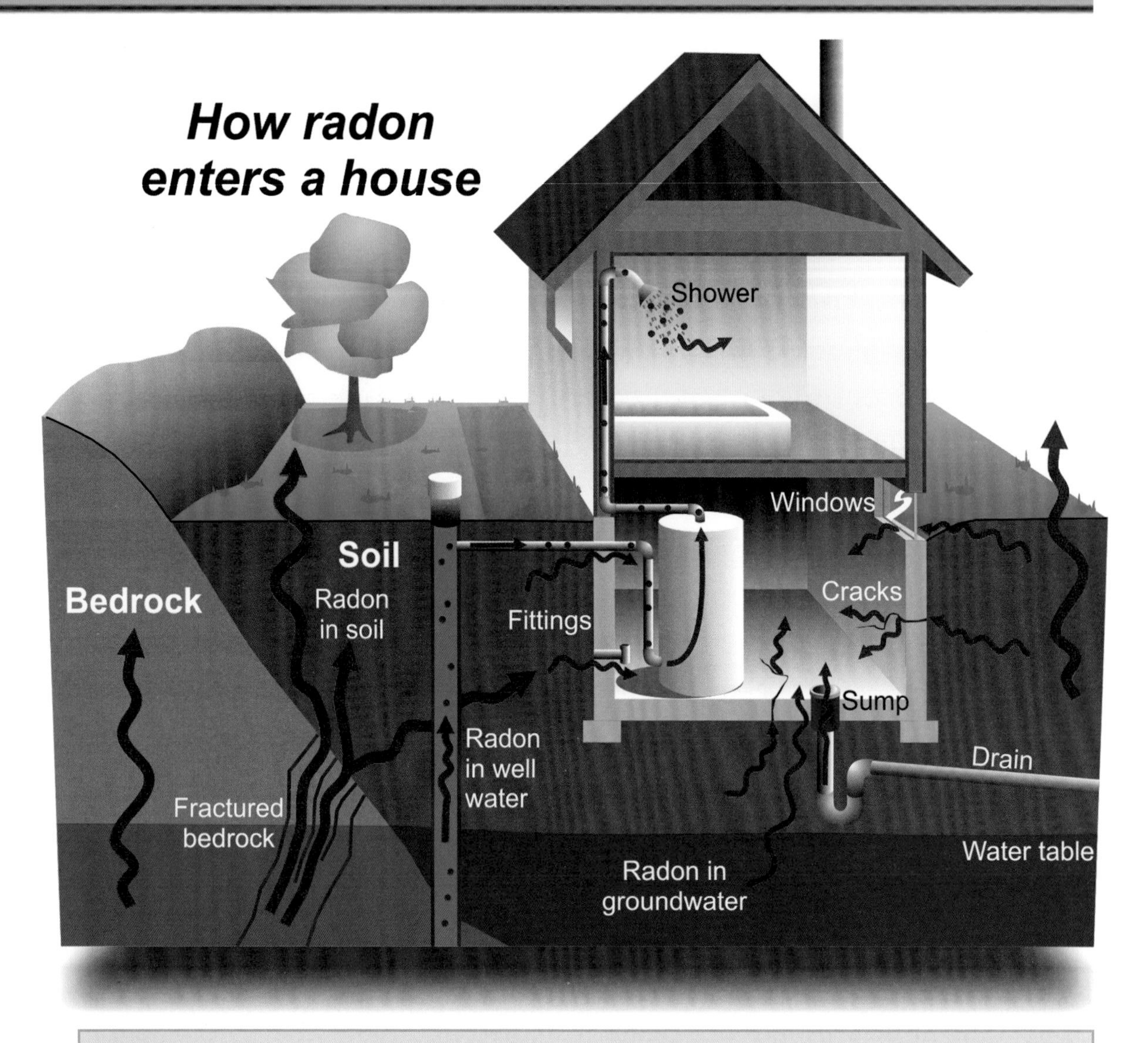

Radon can enter a house many ways. It's important to test your home for radon and take steps to reduce indoor concentrations if they are too high.

occurs in gypsum, a mineral used to make drywall, which is a board used in walls. It also occurs in granite, a hard stone that is often used in public buildings.

Because radon is heavier than air, it tends to collect in basements and on the lower floors of buildings. People living above the third floor of an apartment building are unlikely to have radon problems.

It's easy to find out the radon level in your home. You simply test for it. You can hire a qualified radon professional to do the testing, or you can purchase a do-it-yourself kit. The kit absorbs samples of the air in your home for two or three days. Then you send it to a laboratory for analysis. The laboratory will send the results back to you. According to the U.S. Environmental Protection Agency (EPA), the average indoor air concentration of radon is 1.3 picoCuries per liter (1.3 pCi/l) of air. A Curie is a unit of radioactivity. A picoCurie is one-trillionth of a Curie. The EPA recommends you take steps to lower the radon levels in your home if they exceed 4 pCi/l of air.

Reducing Radon Levels

There are several steps you can take to reduce radon concentrations in your home. An easy step is to seal cracks in basement walls and floors or in the foundation. Sealing cracks in walls and closing the gaps around water and gas pipes also helps. You can even seal cracks in the roof and ceiling. That prevents warm air in the home from rising and leaking out at the top, which draws radon in from below. In addition, sealing air-conditioning and heating ducts in basements or

Reducing radon levels in a home will probably require the services of a professional. The builder kneeling here teaches other builders how to lower indoor radon levels.

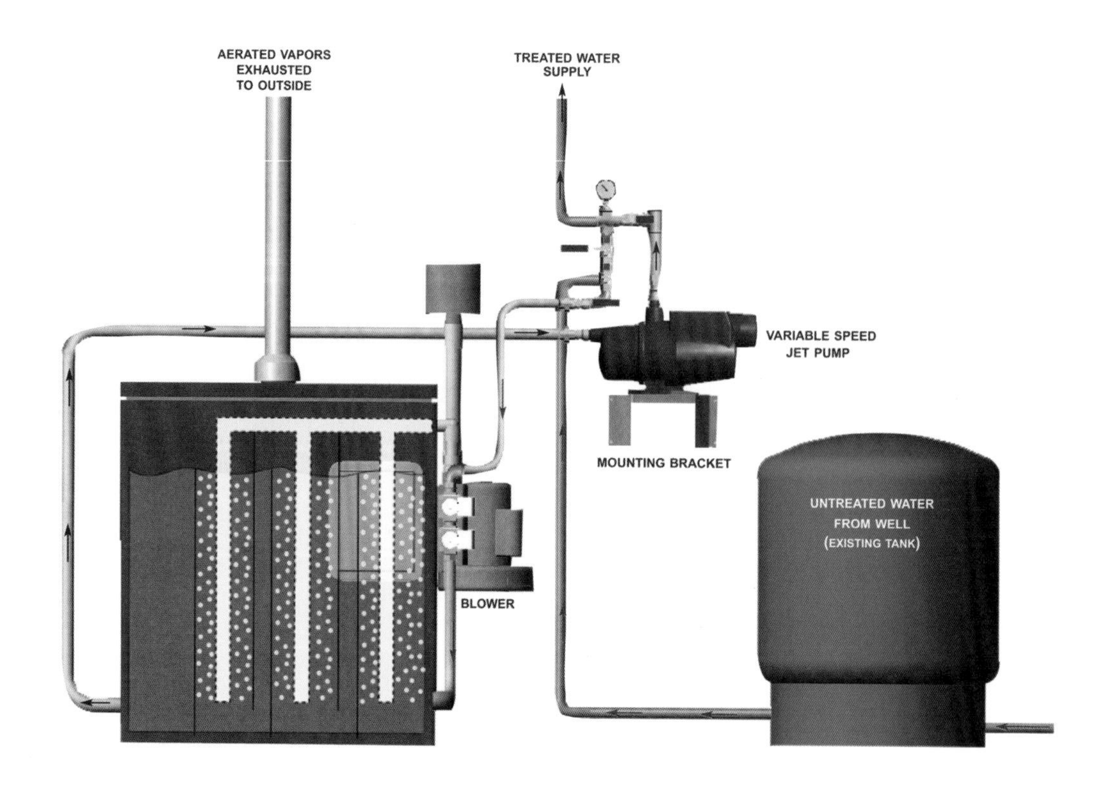

This figure illustrates one type of aerator process, called AlRaider 433, for removing radon from water. A special tank captures radon in the water with air bubbles. The air and radon are vented to the outside.

under houses can prevent them from pulling in radon and distributing it through the house.

Other methods include putting plastic sheeting below the basement or foundation to keep radon from entering the home. You can also create a space below the home where radon can collect instead of entering the home. Radon can be piped from there through the home's walls to the roof and vented above the home, where it poses no danger. If necessary, a fan can be added to help pull the radon up through the pipe.

Devices that remove radon from water can be attached at the point where the water enters your home. One type of device uses a carbon filter to remove radon from the water. Another type bubbles air through

Radon Levels in the Watras Home

Remember Stanley Watras? He was discussed in the introduction of this book. He was a construction engineer whose home turned out to be highly contaminated with radon. Testing showed radon levels in the Watras home were 2,700 pCi/l! The harm to a person's lungs was equivalent to smoking about 200 packs of cigarettes per day or getting 455,000 chest X-rays in one year!

the water. The air bubbles capture the radon and carry it out into the atmosphere through an exhaust fan.

In many places, builders often include radon protection features in new homes. That means it is possible to buy a new house that is resistant to radon. But you don't have to buy a new home to protect yourself from radon. Qualified radon professionals can install the features in existing homes.

Because we know radon can cause lung cancer, which can lead to death, it's important to protect yourself from too much exposure to radon. It may also cause health problems that haven't yet been identified. On the other hand, exposure to small amounts may have health benefits that most scientists don't yet recognize. Perhaps the researchers studying the effects of radon spas and mines will find clear evidence that radon has healing powers. There's much we don't yet know about radon. Future scientists have a lot left to discover.

The Periodic Table of Elements

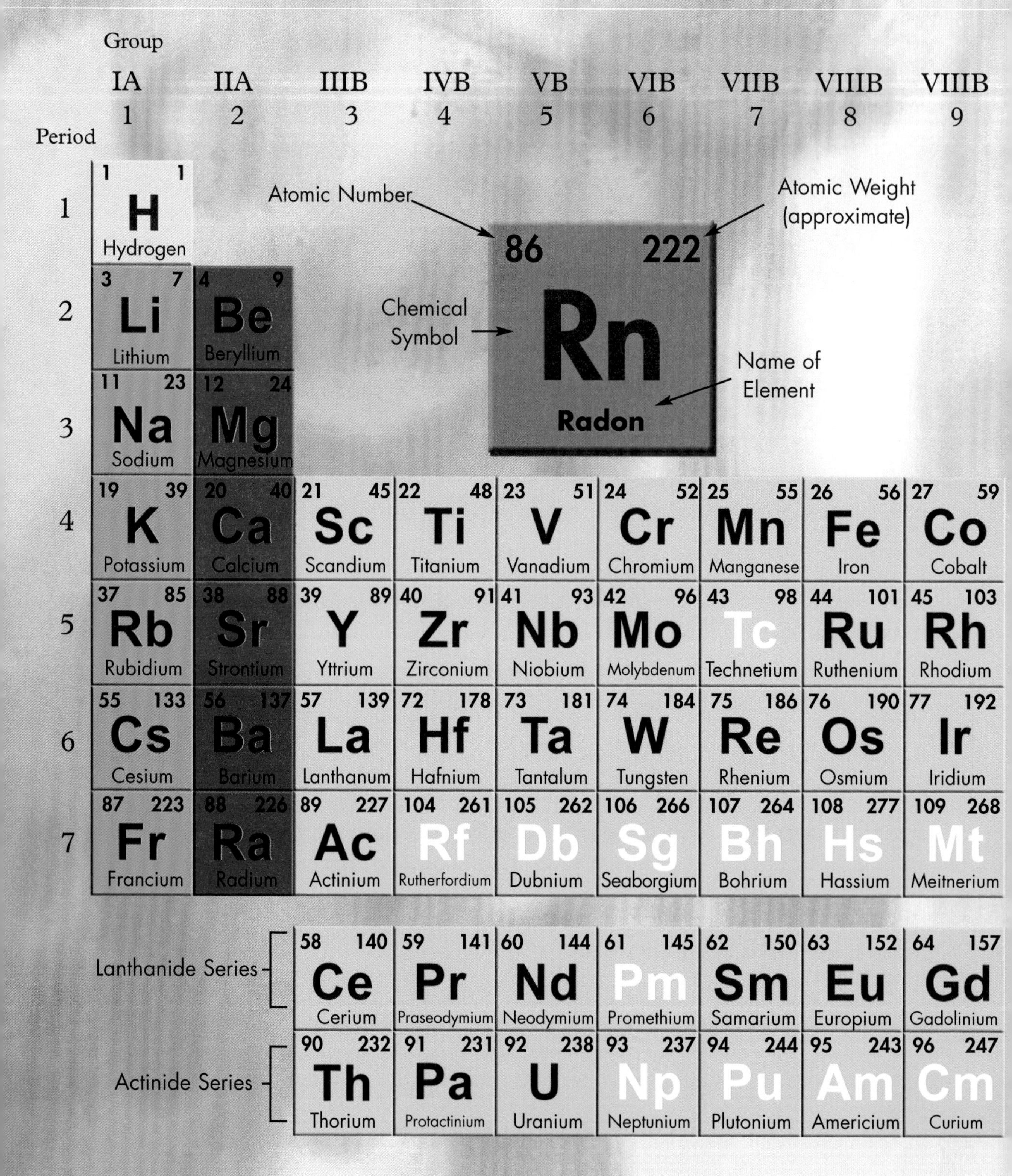

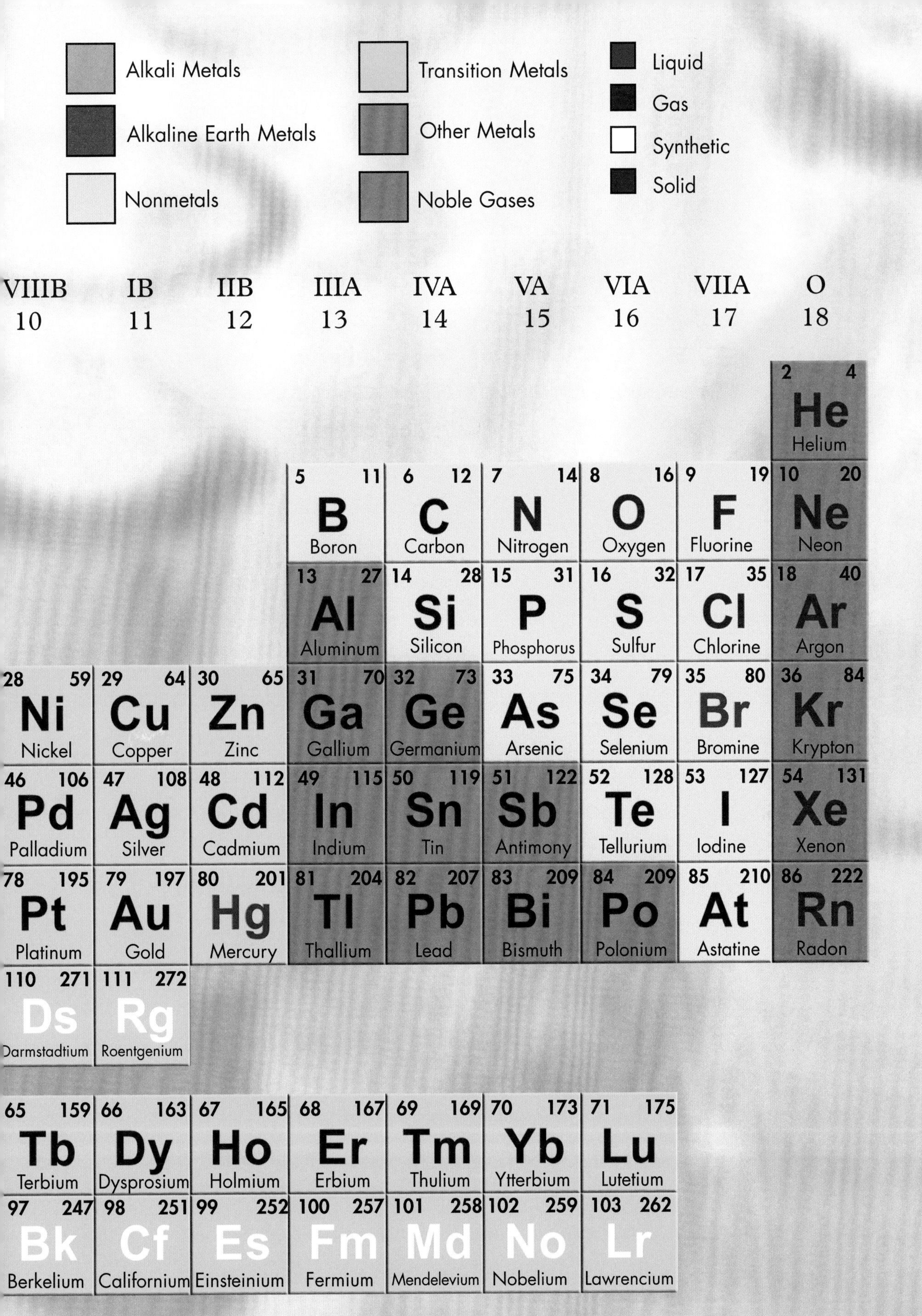
Alkali Metals
Alkaline Earth Metals
Nonmetals
Transition Metals
Other Metals
Noble Gases
Liquid
Gas
Synthetic
Solid
VIIIB 10
IB 11
IIB 12
IIIA 13
IVA 14
VA 15
VIA 16
VIIA 17
O 18
2 4 He Helium
5 11 B Boron
6 12 C Carbon
7 14 N Nitrogen
8 16 O Oxygen
9 19 F Fluorine
10 20 Ne Neon
13 27 Al Aluminum
14 28 Si Silicon
15 31 P Phosphorus
16 32 S Sulfur
17 35 Cl Chlorine
18 40 Ar Argon
28 59 Ni Nickel
29 64 Cu Copper
30 65 Zn Zinc
31 70 Ga Gallium
32 73 Ge Germanium
33 75 As Arsenic
34 79 Se Selenium
35 80 Br Bromine
36 84 Kr Krypton
46 106 Pd Palladium
47 108 Ag Silver
48 112 Cd Cadmium
49 115 In Indium
50 119 Sn Tin
51 122 Sb Antimony
52 128 Te Tellurium
53 127 I Iodine
54 131 Xe Xenon
78 195 Pt Platinum
79 197 Au Gold
80 201 Hg Mercury
81 204 Tl Thallium
82 207 Pb Lead
83 209 Bi Bismuth
84 209 Po Polonium
85 210 At Astatine
86 222 Rn Radon
110 271 Ds Darmstadtium
111 272 Rg Roentgenium
65 159 Tb Terbium
66 163 Dy Dysprosium
67 165 Ho Holmium
68 167 Er Erbium
69 169 Tm Thulium
70 173 Yb Ytterbium
71 175 Lu Lutetium
97 247 Bk Berkelium
98 251 Cf Californium
99 252 Es Einsteinium
100 257 Fm Fermium
101 258 Md Mendelevium
102 259 No Nobelium
103 262 Lr Lawrencium

Glossary

atom The smallest part of an element having the chemical properties of that element.

contamination The state of being made unfit for use.

diffuse To spread out, as when a gas spreads out into the air.

duct A tube, pipe, or channel for carrying heated or cooled air through buildings.

emphysema A lung disease with abnormal expansion of the air spaces and walls of the lungs.

fault A place where two pieces of the earth's crust (tectonic plates) come together.

fluorescent Glowing as a result of absorbing radiation.

half-life The time it takes for half the atoms in a sample of a radioactive element to decay.

inert Not chemically active.

insulated Built with materials to prevent the transfer of heat into or out of a structure.

ion An atom or molecule with an electrical charge, which results from unequal numbers of protons and electrons.

isotopes Atoms containing the same number of protons but different numbers of neutrons.

mineral springs Water that naturally contains mineral salts or gases.

progeny The products of radioactive decay.

property One of the characteristics of an element.

radiation Rays of light, heat, or energy that spread outward from something.

spontaneously Happening without some outside cause.

ventilation A system or means of providing fresh air.

For More Information

Agency for Toxic Substances and Disease Registry (ATSDR)
Department of Health and Human Services
1825 Century Boulevard
Atlanta, GA 30345
(800) 232-4636
Web site: http://www.atsdr.cdc.gov
The ATSDR is a federal public health agency that works to prevent harmful exposures and diseases related to toxic substances by using science, taking public health actions, and providing health information. Radon is among the substances with which the agency is concerned.

American Chemical Society
1155 Sixteenth Street NW
Washington, DC 20036
(800) 227-5558
Web site: http://www.chemistry.org/portal/a/c/s/1/home.html
The American Chemical Society is the national organization for professional chemists. It also provides information about all aspects of chemistry for students and educators.

Radiation Safety Institute of Canada
National Education Centre
1120 Finch Avenue West, Suite 607
Toronto, ON M3J 3H7
Canada
(800) 263-5803
(416) 650-9090

Web site: http://www.radiationsafety.ca
The Radiation Safety Institute of Canada is an independent national organization committed to promoting radiation safety in homes, schools, workplaces, and the environment. Radon is one of the topics with which it's concerned.

U.S. EPA/Office of Radiation and Indoor Air
Indoor Environments Division
1200 Pennsylvania Avenue NW
Mail Code 6609J
Washington, DC 20460
(202) 343-9370
Web site: http://www.epa.gov/iaq/index.html
The mission of the Office of Radiation and Indoor Air is to protect the public and the environment from radiation and indoor air pollution. The issue of radon in homes is one of the problems addressed by the office.

Web Sites

Due to the changing nature of Internet links, Rosen Publishing has developed an online list of Web sites related to the subject of this book. This site is updated regularly. Please use this link to access the list:

http://www.rosenlinks.com/uept/rado

For Further Reading

Diagram Group, The. *The Facts on File Chemistry Handbook*. Rev. ed. New York, NY: Facts on File, 2006.

Emsley, John. *Nature's Building Blocks: An A–Z Guide to the Elements*. New York, NY: Oxford University Press, 2003.

Ganeri, Anita. *Neon and the Noble Gases* (The Periodic Table). Chicago, IL: Heinemann Library, 2004.

Jerome, Kate Boehm. *Atomic Universe: The Quest to Discover Radioactivity* (Science Quest). Washington, DC: National Geographic, 2006.

Karam, P. Andrew, and Ben P. Stein. *Radioactivity*. New York, NY: Chelsea House Publishers, 2008.

Keller, Rebecca W. *Chemistry: Level 1* (Real Science-4-Kids). Albuquerque, NM: Gravitas Publications, Inc., 2005.

Knapp, Brian. *Uranium and Other Radioactive Elements* (Elements). Danbury, CT: Grolier Educational, 2001.

Krull, Kathleen. *Marie Curie* (Giants of Science). New York, NY: Viking Children's Books, 2007.

Miller, Ron. *The Elements: What You Really Want to Know*. Minneapolis, MN: 21st Century, 2006.

Oxlade, Chris. *Elements and Compounds* (Chemicals in Action). Rev. updated ed. Chicago, IL: Heinemann Library, 2007.

Pasachoff, Naomi. *Ernest Rutherford: Father of Nuclear Science* (Great Minds of Science). Berkeley Heights, NJ: Enslow Publishers, 2005.

Tocci, Salvatore. *Hydrogen and the Noble Gases* (A True Book). Danbury, CT: Children's Press, 2004.

Wertheim, Jane. *The Usborne Illustrated Dictionary of Chemistry*. Rev. ed. London, England: Usborne Books, 2008.

Bibliography

ATSDR (Agency for Toxic Substances and Disease Registry). "Toxicological Profile for Radon: Production, Import, Use, and Disposal." Retrieved February 26, 2008 (http://www.atsdr.cdc.gov/toxprofiles/tp145-c4.pdf).

Ball, Philip. *The Ingredients: A Guided Tour of the Elements*. New York, NY: Oxford University Press, 2002.

Ede, Andrew. *The Chemical Element: A Historical Perspective*. Westport, CT: Greenwood Press, 2006.

Environmental Protection Agency. "A Citizen's Guide to Radon: The Guide to Protecting Yourself and Your Family from Radon." November 26, 2007. Retrieved February 18, 2008 (http://epa.gov/radon/pubs/citguide.html).

Environmental Protection Agency. "Radon." November 15, 2007. Retrieved February 19, 2008 (http://www.epa.gov/rpdweb00/radionuclides/radon.html).

Greenwood, N. N., and A. Earnshaw. *Chemistry of the Elements*. 2nd edition. Woburn, MA: Butterworth-Heinemann, 1997.

Marshall, James L., and Virginia R. Marshall. "Ernest Rutherford, the 'True Discoverer' of Radon." *Bulletin for the History of Chemistry*, Vol. 28, No. 2, 2003, pp. 76–83.

Wozniak, Stephen J. *Handbook of Radon: Health, Economic and Building Aspects*. Retrieved February 19, 2008 (http://www.seered.co.uk/radon00.htm).

Yamaoka, Kiyonori, et al. "Biochemical Comparison Between Radon Effects and Thermal Effects on Humans in Radon Hot spring Therapy." *Journal of Radiation Research*, Vol. 45, No. 1, 2004, pp. 83–88.

Index

About the Author

Janey Levy is a writer and editor who has written more than seventy-five books for young people. Her interest in chemistry comes from her curiosity about the role the body's own chemistry plays in an individual's physical, emotional, and mental well-being. She has written about other elements of the periodic table, including krypton and tin. Levy lives in Colden, New York.

Photo Credits

Cover, pp. 1, 14, 16, 19, 20, 40–41 Tahara Anderson; p. 5 © Time & Life Pictures/Getty Images; p. 8 National Library of Medicine/NIH; pp. 9, 25 © Getty Images; pp. 10, 28, 33 USGS; p. 11 © Ria Novosti/Photo Researchers; p. 29 © Will & Demi McIntyre/Photo Researchers; pp. 31, 37 © AP Photos; p. 34 © National Geographic/Getty Images; p. 36 Natural Resources Canada; p. 38 Diagram of AIRaider™ Diffused Bubble Aeration System setup and process courtesy of RadonAway.

Designer: Tahara Anderson; **Editor:** Kathy Kuhtz Campbell
Photo Researcher: Marty Levick